MENTAL MATHS

Book-3

Rajesh Singh

© Young Learner Publications®, India

CONTENTS

Unit	Topic	Worksheet	Page No.
1.	Numbers: 1000 to 10000	1-6	3-8
2.	Operations on Numbers (Addition, Subtraction, Multiplication, Division)	7-23	9-25
3.	Fractions	24-25	26-27
4.	Measurement (Length, Weight, Capacity)	26-37	28-39
5.	Time and Money	38-42	40-44
6.	Data Handling and Patterns	43-45	45-47
7.	Geometry	46-50	48-52
	Answers		53-56

Published by:

 YOUNG LEARNER PUBLICATIONS®

G-1A Rattan Jyoti, 18 Rajendra Place
New Delhi-110008 (INDIA)
Tel.: 011-25750801, 25820556, 25755559
Fax: 91-11-25764396
Email: goodwillpub@gmail.com
gph.ylp@gmail.com
gph.ylp@goodwillpublishinghouse.com
Website: www.goodwillpublishinghouse.com

© Young Learner Publications, India

All rights reserved. No part of this publication may be reproduced, stored in a retrieval system or transmitted, in any form or by any means—mechanical, photocopying, recording or otherwise, without prior written permission of Young Learner Publications, India.

Unit-1 Numbers: 1000 to 10000

Worksheet - 1

Counting

1. Count forward and fill in the boxes.

a) | 2032 | 2033 | 2034 | 2035 | 3036 | 3037 | 3038 | 2039 |

b) | 4158 | 4159 | 4160 | 4161 | 4162 | 4163 | 4164 | 4165 |

c) | 6402 | 6403 | 6404 | 6405 | 8406 | 6407 | 6408 | 6409 | 6410 |

d) | 8365 | 8366 | 8367 | 8368 | 8369 | 8370 | 8371 | 8372 | 8373 |

e) | 9993 | 9994 | 9995 | 9996 | 9997 | 9998 | 9999 | 99910 |

2. Count backward and fill in the boxes.

a) | 1730 | 1729 | 1728 | 1727 | 1726 | 1725 | 1724 | 1723 |

b) | 3224 | 3223 | 3222 | 3221 | 3220 | 3219 | 3218 | 3217 | 3216 |

c) | 5204 | 5203 | 5202 | 5201 | 5200 | 5199 | 5198 |

d) | 7937 | 7936 | 7935 | 7934 | 7933 | 7932 | 7931 |

e) | 9999 | 9998 | 9997 | 9996 | 9995 | 9994 | 9993 |

Worksheet - 2

Number Names and Numerals

1. Write the number name for:

   a) 1121 _____

   b) 2033 _____

   c) 3315 _____

   d) 4477 _____

   e) 5544 _____

   f) 6208 _____

   g) 7766 _____

   h) 8857 _____

   i) 9680 _____

   j) 9999 _____

2. Write the numeral for:

   a) One thousand five hundred eighteen _____

   b) Two thousand eight hundred fifty-one _____

   c) Three thousand four hundred sixty-five _____

   d) Four thousand six hundred twenty-seven _____

   e) Five thousand three hundred seventy-nine _____

   f) Six thousand four hundred thirty-two _____

   g) Seven thousand two hundred eighty-three _____

   h) Eight thousand two hundred forty-six _____

   i) Nine thousand one hundred ninety-four _____

   j) Ten thousand _____

Worksheet - 3

Hundreds, Tens and Ones
After, Before and In Between

1. **Give the numeral.**

   a) 4 thousands 2 hundreds 5 tens 9 ones: _____

   b) 3 thousands 6 hundreds 7 tens 8 ones: _____

2. **Fill in the blanks.**

   a) 2365: _____ thousands _____ hundreds _____ tens _____ ones

   b) Seven thousand four hundred twenty-nine: _____ thousands _____ hundreds _____ tens _____ ones

3. **What comes after?**

   a) 3230, _____   b) 7661, _____, _____

   c) 5697, _____, _____, _____

4. **What comes before?**

   a) _____, 1237   b) _____, _____, 5662

   c) _____, _____, _____, 8000

5. **What comes in between?**

   a) 4122, _____, 4124   b) 5314, _____, _____, 5317

   c) 9996, _____, _____, _____, 10000

6. **What comes before and after?**

   a) _____, 7271, _____

   b) _____, _____, 8363, _____, _____

   c) _____, _____, _____, 9997, _____, _____, _____

Worksheet - 4

Place Value of Digits, Expanded Form
Comparing Numbers, Ascending and Descending Order

Fill in the blanks.

a) 5683: Digit in Th's place _____, digit in H's place _____, digit in T's place _____, digit in O's place _____.

b) The numeral with 6 at thousands place, 7 at hundreds place, 1 at tens place and 0 at ones place is _____.

c) 2317: PV of 2 is _____, PV of 3 is _____, PV of 1 is _____ and PV of 7 is _____.

d) The numeral in which place value of 8 is 8000, 5 is 500, 8 is 80 and 4 is 4 is _____.

e) Expanded form of 8923 is _____ + _____ + _____ + _____.

f) The numeral with expanded form 6000 + 700 + 60 + 3 is _____.

g) Put the correct sign (< or = or >) in the blanks.

   (i) 458 ___ 378   (ii) 527 ___ 527   (iii) 217 ___ 675

   (iv) 4000 + 400 + 40 + 4 ___ 4440   (v) 6051 ___ 6000 + 500 + 1

   (vi) 6 thousands 4 hundreds 2 tens 9 ones ___ 6429

h) An encyclopedia on plants has 5,931 pages and one on animals has 5,082 pages. Which one has more pages? _____

i) Write in ascending order.

   9216, 8329, 8438, 4551   _____

j) Write in descending order.

   4531, 6502, 3343, 1353   _____

Worksheet - 5

Smallest and Greatest Numbers Using Given Digits

1. Make the greatest and smallest 4-digit numbers using 9, 3, 5 and 7. Repetition is not allowed.

   Greatest: _____   Smallest: _____

2. Make the greatest and smallest 4-digit numbers using 4, 2, 6 and 5. Repetition is allowed.

   Greatest: _____   Smallest: _____

3. Make the greatest and smallest 4-digit numbers using 0, 4, 8 and 6. Repetition is not allowed.

   Greatest: _____   Smallest: _____

4. Make the greatest and smallest 4-digit numbers using 0, 2, 5 and 3. Repetition is allowed.

   Greatest: _____   Smallest: _____

5. Make the greatest and smallest 4-digit numbers using four digits out of 1, 2, 3, 4 and 7. Repetition is not allowed.

   Greatest: _____   Smallest: _____

6. Make the greatest and smallest 4-digit numbers using four digits out of 4, 6, 7, 8 and 9. Repetition is allowed.

   Greatest: _____   Smallest: _____

7. Make the greatest and smallest 4-digit numbers using four digits out of 9, 8, 6, 5 and 0. Repetition is not allowed.

   Greatest: _____   Smallest: _____

8. Make the greatest and smallest 4-digit numbers using four digits out of 0, 2, 5, 7 and 9. Repetition is allowed.

   Greatest: _____   Smallest: _____

Worksheet - 6

Rounding-off Numbers

1. **Fill in the blanks.**

   To round off to nearest ten we look at the digit in _____ place. If this is less than 5 we write the digit in O's place as _____. If this is 5 or more we write the digit in O's place as _____ and add 1 to _____ digit.

2. **Round off to nearest ten.**

   a) 4825 ≈ _____   b) 5307 ≈ _____   c) 6523 ≈ _____

3. **Fill in the blanks.**

   To round off to nearest hundred we look at the digit in _____ place. If this is less than 5 we write the digits in O's place and T's place as _____. If this is 5 or more we write the digit in O's place and T's place as _____ and add 1 to _____ digit.

4. **Round off to nearest hundred.**

   a) 5171 ≈ _____   b) 4259 ≈ _____   c) 6625 ≈ _____

5. **Fill in the blanks.**

   To round off to nearest thousand we look at the digit in _____ place. If this is less than 5 we write the digits in O's place, T's place and H's place as _____. If this is 5 or more we write the digit in O's place, T's place and H's place as _____ and add 1 to _____ digit.

6. **Round off to nearest thousand.**

   a) 2495 ≈ _____   b) 6857 ≈ _____   c) 4523 ≈ _____

7. **A book fair was visited by 4,385 people. The number of visitors,**

   a) rounded off to nearest ten is _____.

   b) rounded off to nearest hundred is _____.

   c) rounded off to nearest thousand is _____.

   d) Which is the most accurate? _____

Note: ≈ denotes rounding off.

Unit-2 Operations on Numbers

Worksheet - 7

Addition Without Carrying and With Carrying

1. Addition with and without carrying.

a)
```
  6 3 2 5
+ 2 6 3 2
---------
  8 9 5 7
```

b)
```
  2 1 6 2
+ 4 7 2 3
---------
  6 8 8 5
```

c)
```
  7 4 2 4
+ 1 2 5 4
---------
  8 6 7 8
```

d)
```
  6 2 5 6
+ 2 4 3 2
---------
  8 6 8 8
```

e) (carries: 1 1 1)
```
  6 7 9 4
+ 1 2 3 7
---------
  8 0 3 2
```

f) (carries: 1 1 1)
```
  4 5 9 6
+ 2 3 6 5
---------
  6 9 6 1
```

g) (carries: 1 1 1)
```
  5 9 8 5
+ 1 6 7 8
---------
  7 6 6 3
```

h) (carries: 1 1 1)
```
  4 8 8 9
+ 2 4 2 3
---------
  7 3 1 2
```

i)
```
  4 1 0 4
  3 4 5 3
+ 2 2 2 1
```

j)
```
  4 5 4 6
  2 2 4 1
+ 2 2 0 0
```

k)
```
  5 3 0 1
  2 4 3 4
+     3 2
```

l)
```
  6 2 1 2
      1 2 1
+   5 6 3
```

m)
```
  4 1 8 4
  1 4 5 9
+ 1 2 2 1
```

n)
```
  2 5 4 6
  3 5 8 6
+ 2 2 0 9
```

o)
```
  4 3 0 1
  2 9 3 3
+     9 8
```

p)
```
      8 2
  6 2 3 2
+   6 6 8
```

2. Solve the following:

a) 4427 + 3351 = 7778

b) 6294 + 3001 = 9295

c) 3907 + 2275 = 6182

d) 3986 + 4194 = 7080

e) 2201 + 2203 + 1312 = _____

f) 1333 + 4222 + 2500 = _____

Worksheet - 8

Addition Stories

1. 3,725 people visited a shrine yesterday and 4,908 people visited it today. How many people visited the shrine in these two days?
   _____
   _____

2. A dealer sold 3,132 scooters in the first year and 3,034 scooters in the second year. How many scooters did the dealer sell in these two years?
   _____
   _____

3. A colony has two sectors A and B. 1,342 people live in Sector A and 2,259 people live in Sector B. How many people live in the colony?
   _____
   _____

4. In a school festival 2,093 tickets were sold on the first day. On the second day same number of tickets were sold. How many tickets were sold in the two days?
   _____
   _____

5. A factory made 1,520 items in January, 1,798 items in February and 2,507 items in March. How many items did it make in these three months?
   _____
   _____

6. A factory has 1,261 workers. Another factory has 1,762 workers. How many workers are working in the two factories?
   _____
   _____

Worksheet - 9

Estimating Sum and Addition Facts

1. Estimate by rounding to nearest ten.

   a) 2137 + 3258 ≈ _____ + _____ = _____

   b) 2085 + 3818 ≈ _____ + _____ = _____

2. A box has 2,612 items. Another box has 5,723 items. Estimate the total number of items in both boxes by rounding to nearest ten.

   _____ + _____ ≈ _____ + _____ = _____

3. Estimate by rounding to nearest hundred.

   a) 2135 + 3258 ≈ _____ + _____ = _____

   b) 4612 + 2834 ≈ _____ + _____ = _____

4. A bag has 4,566 pens. Another bag has 4,273 pens. Estimate the total number of pens in the two bags by rounding to nearest hundred.

   _____ + _____ ≈ _____ + _____ = _____

5. Estimate by rounding to nearest thousand.

   a) 4238 + 5417 ≈ _____ + _____ = _____

   b) 3987 + 4195 ≈ _____ + _____ = _____

6. An exhibition had 4,726 visitors on Saturday and 3,959 visitors on Monday. Estimate the total number of visitors in the two days by rounding to nearest thousand.

   _____ + _____ ≈ _____ + _____ = _____

7. Fill in the blanks.

   a) _____ + 4132 = 4132 + 7123

   b) _____ + 4623 = _____ + 3874

   c) [2598 + 1194] + _____ = 2598 + [1194 + 2655]

   d) [_____ + 3178] + _____ = 2563 + [_____ + 4359]

   e) 6377 + _____ = 6377

   f) 0 + 4295 = _____

   g) _____ + 1297 = 1297

Worksheet - 10

Subtraction Without Borrowing and With Borrowing

1. Subtraction with and without borrowing.

a) 7754 − 5342 = 2412

b) 8784 − 5463 = 3321

c) 8576 − 4243 = 4333

d) 9896 − 5331 = 4555

e) 4645 − 2331 = 2314

f) 9783 − 7743 = 2040

g) 7996 − 4654 = 3342

h) 8687 − 5432 = 3255

i) 5440 − 4213 = 227

j) 6464 − 3273 = 3191

k) 5569 − 2868 = 2701

l) 8376 − 5289 = 3087

m) 7621 − 2734 = 4887

n) 9065 − 7777 = 1288

o) 7006 − 4877 = 2129

p) 8000 − 5432 = 2558

2. Solve the following:

a) 8877 − 5120 = 3857

b) 8973 − 3322 = 5951

c) 4573 − 3451 = 1122

d) 8636 − 5040 = 3596

e) 6785 − 4421 = 2364

f) 9787 − 2131 = _____

g) 5923 − 4034 = 1889

h) 9900 − 2311 = _____

Worksheet - 11

Addition With Subtraction

Solve the following:

1) 2347 + 2391 − 4698

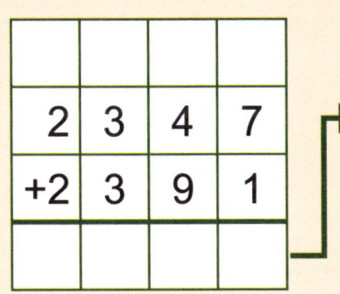
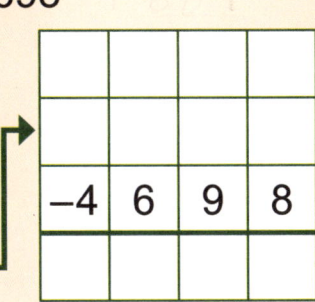

2) 6752 − 2343 + 1412

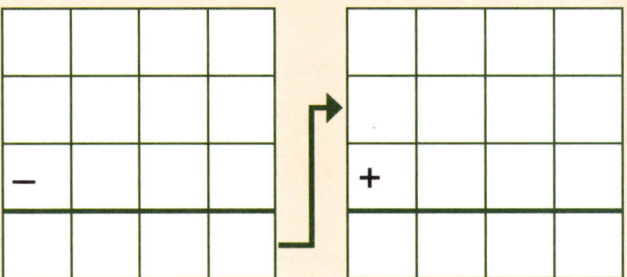

3) 3451 + 2148 − 1286

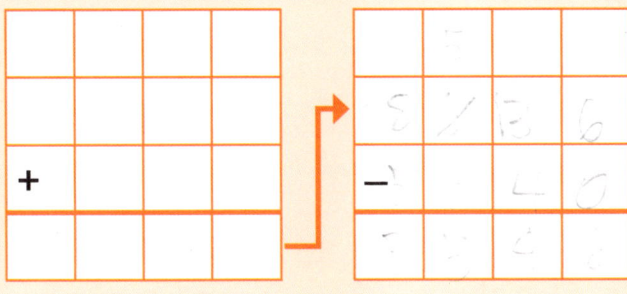

4) 3492 − 2223 + 1374

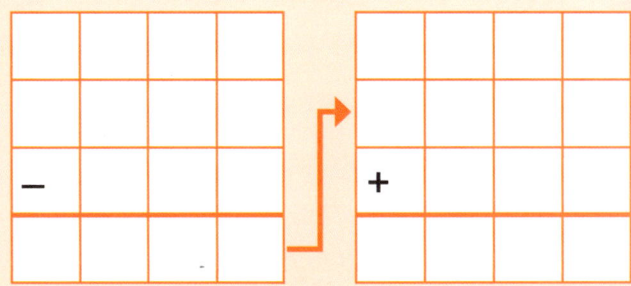

5) 1254 + 3445 − 3034

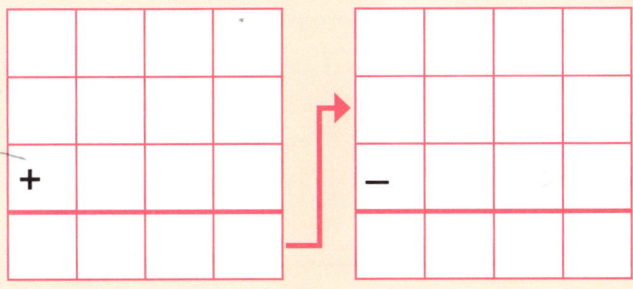

6) 5656 − 2353 + 1214

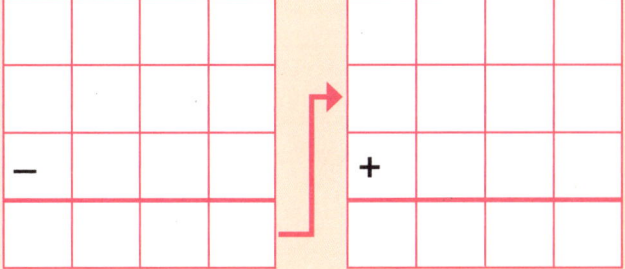

7) 2276 + 1289 − 1620

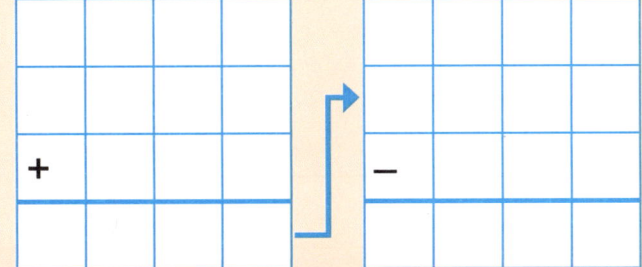

8) 4827 − 1741 + 2691

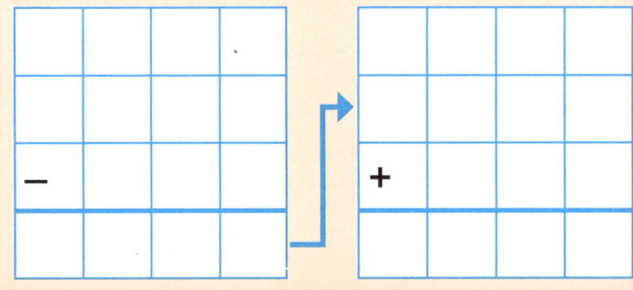

Worksheet - 12

Subtraction Stories

1. To construct a wall 7,500 bricks were bought. After completion of the wall, 2,265 bricks were left. How many bricks were used?

2. There were 3,825 birds in a sanctuary. If 1,621 birds migrate, how many birds are left?

3. There are 2,447 boys and 2,615 girls in a school. How many more girls than boys study in this school?

4. A godown had 3,440 bags of cement. Of this 2,357 bags were sold. How many bags are left in the godown now?

5. Sam has 2,389 coins in his collection. Jane has 1,725 coins in her collection. Who has more coins and by how much?

6. In a warehouse there are 4,961 boxes of shoes. Out of these 2,671 are sent to the market. How many boxes of shoes are left in the warehouse?

Worksheet - 13

Estimating Difference

1. Estimate by rounding to nearest ten.

   a) 9941 − 5457 ≈ _____ − _____ = _____

   b) 7137 − 4259 ≈ _____ − _____ = _____

2. A bag has 6,788 hairpins. Another bag has 4,965 hairpins. Estimate the difference between the number of hairpins by rounding to nearest ten.

   _____ − _____ ≈ _____ − _____ = _____

3. Estimate by rounding to nearest hundred.

   a) 8676 − 5892 ≈ _____ − _____ = _____

   b) 7135 − 4256 ≈ _____ − _____ = _____

4. A college had 4,523 students last year. This year it has 5,812 students. Estimate the number of new students who joined the college this year by rounding to nearest hundred.

   _____ − _____ ≈ _____ − _____ = _____

5. Estimate by rounding to nearest thousand.

   a) 7192 − 3537 ≈ _____ − _____ = _____

   b) 8535 − 4922 ≈ _____ − _____ = _____

6. A packet has 7,655 buttons. Another packet has 5,572 buttons. Estimate the difference between the number of buttons by rounding to nearest thousand.

   _____ − _____ ≈ _____ − _____ = _____

Worksheet - 14

Checking Addition and Subtraction Subtraction Facts

1. **Check the following additions by subtraction:**

   If sum – any one addend = other addend, then subtraction is correct.

   a)
   ```
     2 4 1 8        5 8 7 5
   + 3 4 7 7      – 3 4 7 7
   ─────────      ─────────
     5 8 7 5        _____
   ```
   Addition is _____. (correct/incorrect)

   b)
   ```
     4 7 8 8        _____
   + 3 4 5 5      –  
   ─────────      ─────────
     8 2 4 3        _____
   ```
   Addition is _____. (correct/incorrect)

2. **Check the following subtractions by addition:**

   If subtrahend + difference = minuend, then addition is correct.

   a)
   ```
     6 2 8 7        3 0 5 4
   – 3 0 5 4      + 3 2 3 3
   ─────────      ─────────
     3 2 3 3        _____
   ```
   Subtraction is _____. (correct/incorrect)

   b)
   ```
     4 3 6 7        _____
   – 2 8 2 4      +  
   ─────────      ─────────
     1 4 4 3        _____
   ```
   Subtraction is _____. (correct/incorrect)

3. **Fill in the blanks.**

   a) _____ – 0 = 3424   b) 5198 – _____ = 5198
   c) 3952 – 0 = _____   d) _____ – 2475 = 0
   e) 3677 – _____ = 0   f) 4232 – 4232 = _____

Worksheet - 15

Tables 11 to 15

1. Fill in the boxes to complete the multiplication grid.

×	11	12	13	14	15
1					
2					
3					
4					
5					
6					
7					
8					
9					
10					

2. Each carton has 8 bundles. How many bundles will 11 cartons have?

   Number of bundles = _____ × _____ = _____

3. Each packet has 4 toys. How many toys will 12 packets have?

   Number of toys = _____ × _____ = _____

4. Each van can carry 9 passengers. How many passengers can 13 vans carry?

   Number of passengers = _____ × _____ = _____

5. Each pouch has 6 pencils. How many pencils will 14 pouches have?

   Number of pencils = _____ × _____ = _____

6. Each team has 7 boys. How many boys will 15 teams have?

   Number of boys = _____ × _____ = _____

Worksheet - 16

Multiplication of 2, 3 and 4 Digit Numbers

Solve the following:

1) 4267 × 2

2) 3096 × 3

3) 2138 × 4

4) 1564 × 5

5) 1223 × 6

6) 1237 × 7

7) 1147 × 8

8) 1053 × 9

9) 573 × 12

10) 454 × 13

11) 544 × 14

12) 537 × 15

13) 618 × 15

14) 493 × 14

15) 684 × 13

16) 587 × 12

Worksheet - 17

Multiplication Stories

1. Each basket has 235 toffees. How many toffees will 11 such baskets have?
   _____
   _____

2. A packet has 625 screws. How many screws will 12 such packets have?
   _____
   _____

3. A pouch has 376 key rings. How many key rings will 13 such pouches have?
   _____
   _____

4. To build a section of a boundary wall 572 bricks are required. How many bricks are required to make 14 such sections?
   _____
   _____

5. A truck can carry 585 bags in one round. How many such bags will it carry in 15 rounds?
   _____
   _____

6. A bag has 40 chocolates. How many chocolates will 12 such bags have?
   _____
   _____

Worksheet - 18

Multiplication Facts
Multiplying by 10, 100 and 1000

Fill in the blanks.

a) 4312 × 3119 = 3119 × _____

b) _____ × 8136 = 8136 × 2108

c) 5395 × _____ = 1232 × _____

d) [_____ × 3427] × 5162 = 3982 × [3427 × 5162]

e) [1528 × 3642] × _____ = 1528 × [_____ × 4991]

f) [_____ × 5126] × 3563 = 4139 × [_____ × _____]

g) _____ × 3725 = 3725

h) 2616 × 1 = _____

i) 3396 × _____ = 3396

j) 1 × 5116 = _____

k) _____ × 9619 = 0

l) 3796 × 0 = _____

m) 0 × 43 × 126 = _____

n) 143 × 12 × 1 × 0 = _____

o) 457 × 10 = _____

p) 938 × 10 = _____

q) Each packet has 10 spoons. How many spoons will 27 packets have? _____

r) 27 × 100 = _____

s) 72 × 100 = _____

t) A box has 42 pouches. Each pouch has 100 pens. How many pens in all does the box have? _____

u) 2 × 1000 = _____

v) 7 × 1000 = _____

w) A ship can carry 1,000 passengers in a trip. How many passengers will it carry in 8 trips? _____

Worksheet - 19

Multiplying by Multiples of 10 and Estimating Product

1. Fill in the blanks.

   a) 234 × 20 = _____

   b) 30 × 132 = _____

   c) Each crate has 40 eggs. How many eggs will 224 crates have? _____

   d) 18 × 400 = _____

   e) 500 × 19 = _____

   f) A ferry can transport 200 passengers in a trip. How many passengers will it transport in 37 trips? _____

   g) 2 × 4000 = _____

   h) 2 × 3000 = _____

   i) A truck can carry 3,000 crates of apples in one round. How many crates will it carry in 3 rounds? _____

   j) 80 × 60 = _____

   k) 70 × 30 = _____

   l) Each book has 40 pages. How many pages will 80 copies of this book have? _____

   m) 400 × 20 = _____

   n) 30 × 200 = _____

   o) Each bag has 200 items. How many items will 40 such bags have? _____

   p) 40 × 210 = _____

   q) 30 × 320 = _____

   r) Each bag has 230 items. How many items will 30 such bags have? _____

2. Estimate the product by rounding to nearest tens.

   a) 63 × 45   ≈ _____ × _____ = _____

   b) 127 × 31   ≈ _____ × _____ = _____

   c) Each packet has 39 items. Estimate the number of items contained in 97 such packets by rounding to nearest tens.

   _____ × _____ ≈ _____ × _____ = _____

Worksheet - 20

Division of 2 and 3 Digit Numbers

Solve the following:

1) 11)66

Quotient is _____.

2) 12)96

Quotient is _____.

3) 13)65

Quotient is _____.

4) 14)672

Quotient is _____.

5) 15)930

Quotient is _____.

6) 11)428

Quotient is _____.
Remainder is _____.

7) 13)969

Quotient is _____.
Remainder is _____.

8) 14)609

Quotient is _____.
Remainder is _____.

9) 15)956

Quotient is _____.
Remainder is _____.

Worksheet - 21

Division of 4 Digit Numbers

Solve the following:

1) 11)7139

Quotient is _____.

2) 12)8784

Quotient is _____.

3) 13)6773

Quotient is _____.

4) 14)4984

Quotient is _____.

5) 15)4875

Quotient is _____.

6) 11)6782

Quotient is _____.
Remainder is _____.

7) 12)4995

Quotient is _____.
Remainder is _____.

8) 14)3239

Quotient is _____.
Remainder is _____.

9) 15)7096

Quotient is _____.
Remainder is _____.

Worksheet - 22

Division Stories

1) A tray had 84 eggs. They were packed into packets of 12 eggs each. How many packets were made?

_____ packets were made.

2) A carton has 965 pencil boxes. They have to be packed into packets of 13 pencil boxes each. How many packets can be made? How many pencil boxes will be left behind?

_____ packets can be made.
_____ pencil boxes will be left behind.

3) 5,096 balls are to be packed into packets of 14 balls each. How many packets can be made?

_____ packets can be made.

4) 7,240 combs are to be packed into packets of 15 combs each. How many packets can be made? How many combs will be left behind?

_____ packets can be made.
_____ combs will be left behind.

Worksheet - 23

Division by 10, 100 and 1000
Division Facts, Remainder on Division by 10

1. **Divide by 10, 100, 1000 without actual division.**

   a) 3470 ÷ 10 = _____

   b) 7340 ÷ 10 = _____

   c) 6,750 items were divided into 10 equal parts. How many items will each part have? _____

   d) 3400 ÷ 100 = _____

   e) 4600 ÷ 100 = _____

   f) 5,100 pens were packed into packets of 100 pens each. How many packets were made? _____

   g) 5000 ÷ 1000 = _____

   h) 8000 ÷ 1000 = _____

   i) 9,000 spoons were packed into packets of 1,000 spoons each. How many packets were made? _____

2. **Fill in the blanks.**

   a) 5613 ÷ 5613 = _____

   b) 1568 ÷ _____ = 1

   c) _____ ÷ 6393 = 1

   d) 0 ÷ 3124 = _____

   e) _____ ÷ 3771 = 0

   f) Given that 4572 ÷ 12 = 381. Then 4572 = _____ × _____

   g) Given that 6773 ÷ 13 = 521. Then _____ = 521 × _____

3. **Without actual division find the remainder when each of the following is divided by 10.**

   a) 4237, remainder = _____

   b) 1623, remainder = _____

Worksheet - 24

Express the Fraction

1. Shade $\frac{1}{2}$ of each.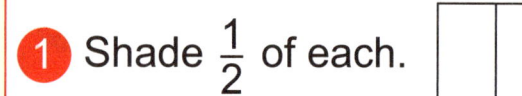

2. Shade $\frac{1}{3}$ of each.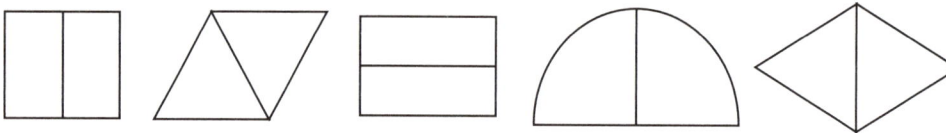

3. Shade $\frac{2}{3}$ of each.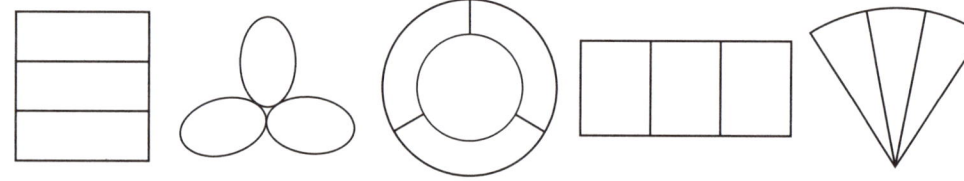

4. Shade $\frac{1}{4}$ of each.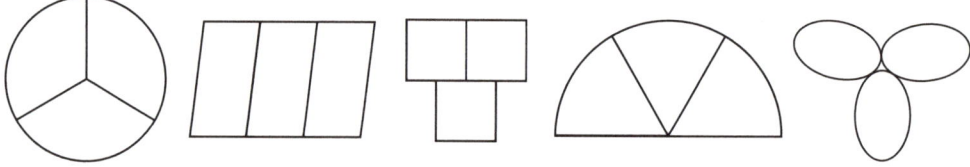

5. Shade $\frac{3}{4}$ of each.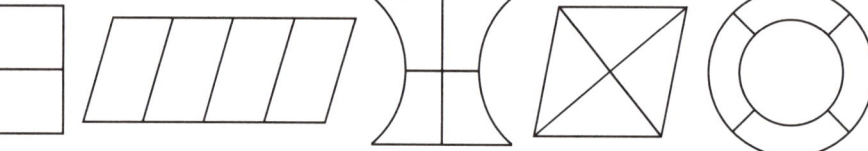

6. Shade $\frac{3}{5}$ of each.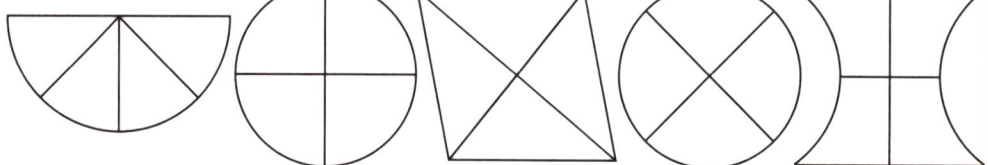

7. Shade $\frac{5}{6}$ of each.

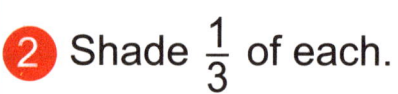
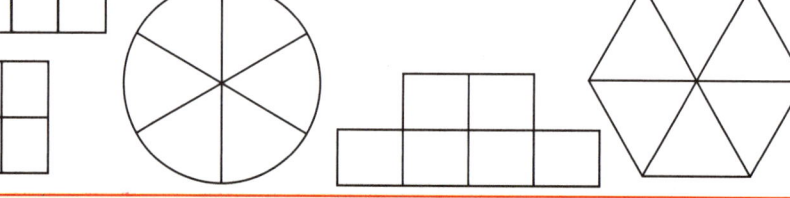

Worksheet - 25

Read the Fraction

What fraction does the shaded part represent in each of the following figures? Write your answer in the box provided.

1	2	3
4	5	6
7	8	9
10	11	12

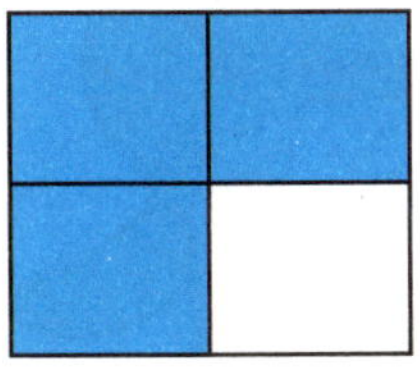

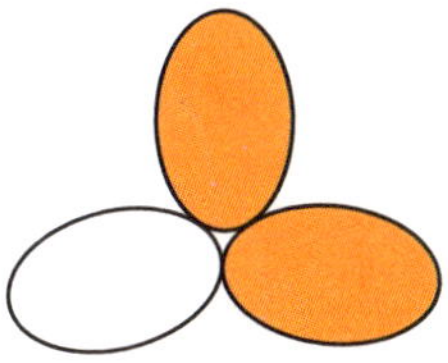

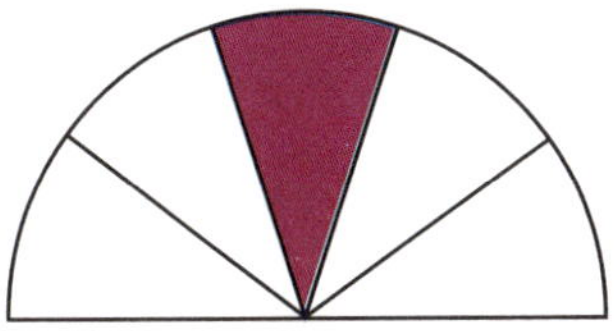

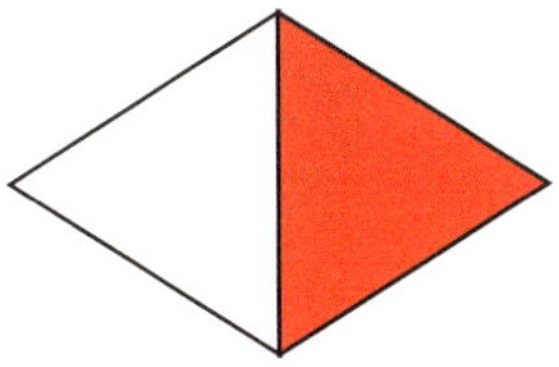

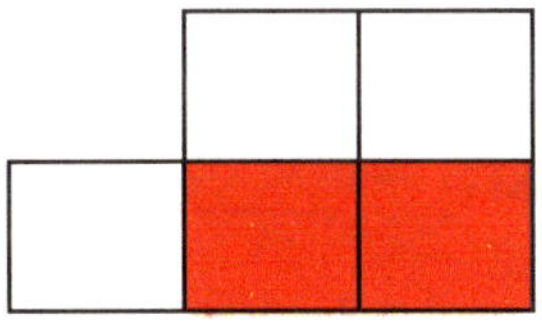

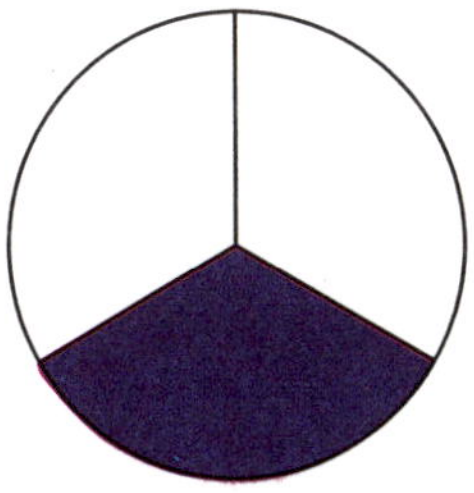

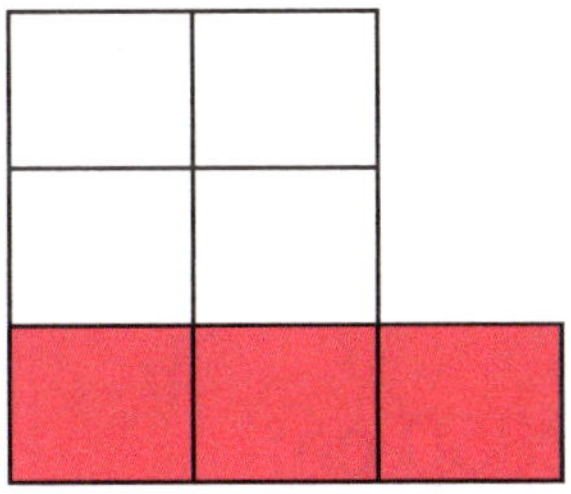

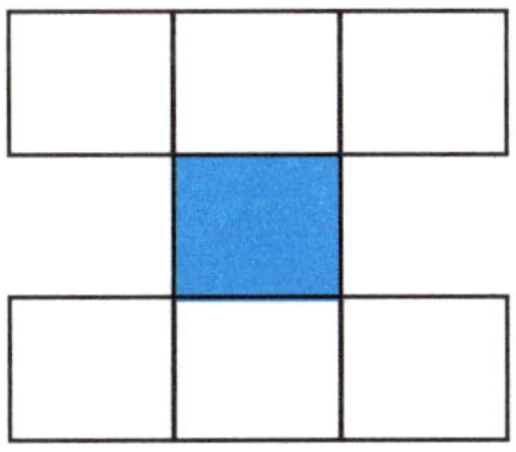

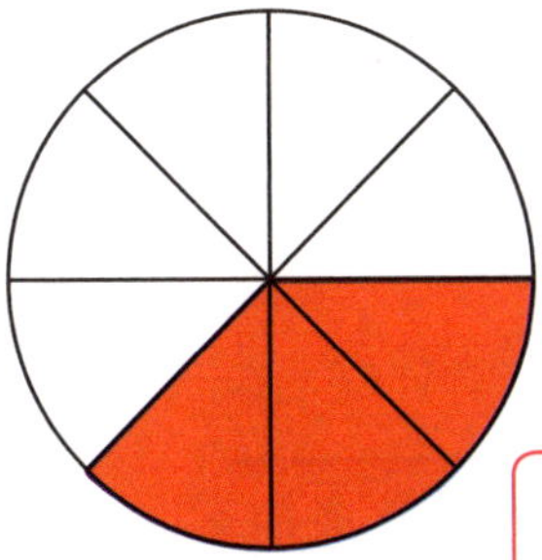

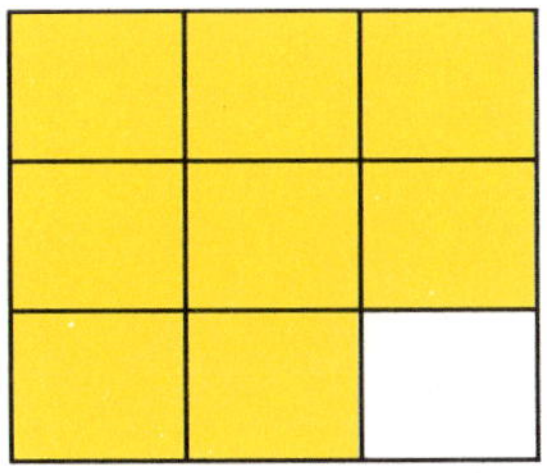

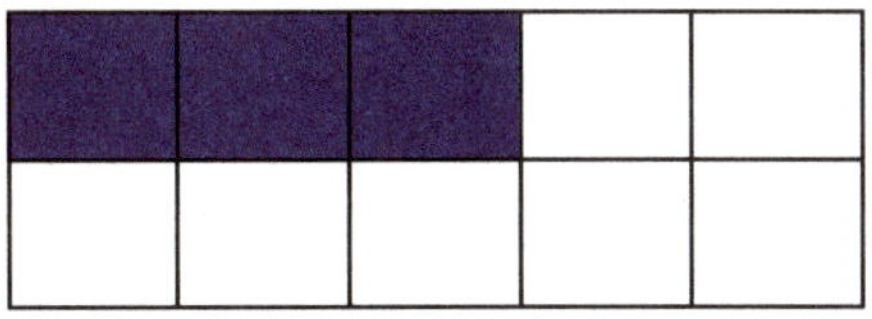

Worksheet - 26

Converting Units of Length

Fill in the blanks.

a) 1 m = _____ cm

b) 4 m = _____ cm

c) 12 m = _____ cm

d) 100 cm = _____ m

e) 400 cm = _____ m

f) 3400 cm = _____ m

g) 1 km = _____ m

h) 8 km = _____ m

i) 7 km = _____ m

j) 1000 m = _____ km

k) 2000 m = _____ km

l) 5000 m = _____ km

m) 17 m 15 cm = _____ cm

n) _____ m _____ cm = 2950 cm

o) A wire is 12 m and 70 cm long. Its length in cm is _____.

p) A chain is 6725 cm long. It is _____ m and _____ cm long.

q) 7 km 200 m = _____ m

r) _____ km _____ m = 9400 m

s) A road is 7 km and 625 m long. Its length in metres is _____.

t) A road is 5720 m long. Its length is _____ km and _____ m.

Worksheet - 27

Addition, Subtraction, Multiplication and Division of Lengths

1. Add the following:

a	b	c
3 4 cm + 6 4 cm ——————— _____ cm	2 4 8 m + 5 3 4 m ——————— _____ m	3 7 9 1 km 4 2 6 4 km + 1 9 3 9 km ——————— _____ km

2. Subtract the following:

a	b	c
7 6 cm − 2 1 cm ——————— _____ cm	9 8 7 m − 3 9 1 m ——————— _____ m	6 3 5 7 km − 4 5 7 5 km ——————— _____ km

3. Multiply the following:

a	b	c
7 3 cm × 1 1 ——————— _____ _____ _____ cm	1 4 5 m × 1 2 ——————— _____ _____ _____ m	4 1 2 km × 1 3 ——————— _____ _____ _____ km

4. Divide the following:

a	b	c
374 cm by 11	2198 m by 14	8805 km by 15

Worksheet - 28

Word Problems

1. A path was 363 m long. It was extended at one end by 214 m. How long is it now?

2. A wire was 1281 m long. A 149 m long piece was cut out of it. What is the length of the remaining wire?

3. A cable was cut into 12 pieces of equal length. What was the original length of the cable if each piece is 231 m long?

4. A rope was cut into 14 pieces of equal length. What is the length of each piece if the rope was 95 m 48 cm long?

5. 13 flagpoles have been erected at equal distances along the side of a road. If the distance between the first and the last flagpole is 4824 m, what is the distance between any two successive poles?

6. A wire is 20 m long. Another wire of length 81 m is joined to it. What is the total length of the wire now?

Worksheet - 29

Estimating Length

1. **Estimate the following by rounding to nearest 10 cm:**

    a) 83 cm ≈ _____ cm   b) 135 cm ≈ _____ cm

2. **Estimate the following by rounding to nearest 100 cm:**

    a) 621 cm ≈ _____ cm = _____ m

    b) 898 cm ≈ _____ cm = _____ m

3. **Estimate the following by rounding to nearest 10 m:**

    a) 52 m ≈ _____ m   b) 175 m ≈ _____ m

4. **Estimate the following by rounding to nearest 100 m:**

    a) 921 m ≈ _____ m   b) 7386 m ≈ _____ m

5. **Estimate the following by rounding to nearest 1000 m:**

    a) 7432 m ≈ _____ m = _____ km

    b) 2595 m ≈ _____ m = _____ km

6. **Estimate the following by rounding to nearest 10 km:**

    a) 896 km ≈ _____ km   b) 1214 km ≈ _____ km

7. **Estimate the following by rounding to nearest 100 km:**

    a) 867 km ≈ _____ km   b) 2439 km ≈ _____ km

8. **Estimate the following by rounding to nearest 1000 km:**

    a) 3162 km ≈ _____ km   b) 6825 km ≈ _____ km

Worksheet - 30

Converting Units of Weight

Fill in the blanks.

a) 1 kg = _____ g

b) To convert kg to g we _____ by _____.

c) 4 kg = _____ g

d) 8 kg = _____ g

e) A suitcase weighs 5 kg. Its weight in g is _____.

f) To convert 7 kg to g we can either multiply by _____ or simply put _____ zeros in the end.

g) 1000 g = _____ kg

h) To convert g to kg we _____ by _____.

i) 3000 g = _____ kg

j) 8000 g = _____ kg

k) A packet weighs 7000 g. Its weight in kg is _____.

l) To convert 4000 g to kg we can either divide by _____ or simply remove _____ zeros in the end.

m) A parcel weighs 6 kg and 250 g. Its weight in g is _____.

n) A school bag weighs 3 kg and _____ g. Its weight in g is 3560.

o) A pouch weighs _____ kg and 225 g. Its weight in g is 4225.

p) A box weighs 4510 g. Its weight in kg is _____ and in g is _____.

q) A carton weighs 2 kg and 350 g. Its weight in g is _____.

r) A bundle weighs _____ kg and 650 g. Its weight in g is 7650.

s) A carton weighs 3 kg and _____ g. Its weight in g is 3720.

t) A baby weighs 6830 g. His weight is _____ kg and _____ g.

Worksheet - 31

Adding, Subtracting, Multiplying and Dividing Weights

1. Add the following:

a)
```
   3 6 g
 + 4 2 g
 ───────
       g
```

b)
```
   7 4 8 g
 + 1 3 9 g
 ─────────
         g
```

c)
```
   6 5 1 8 kg
 + 1 7 1 3 kg
 ────────────
           kg
```

2. Subtract the following:

a)
```
   9 6 g
 − 8 4 g
 ───────
       g
```

b)
```
   8 0 6 3 kg
 − 1 4 0 1 kg
 ────────────
           kg
```

c)
```
   2 7 8 5 kg
 −   2 1 7 kg
 ────────────
           kg
```

3. Multiply the following:

a)
```
   3 8 9 g
 ×   1 2
 ───────
 ───────
 ───────
       g
```

b)
```
   2 1 6 kg
 ×     1 3
 ─────────
 ─────────
 ─────────
        kg
```

c)
```
   4 3 4 kg
 ×     1 4
 ─────────
 ─────────
 ─────────
        kg
```

4. Divide the following:

a) 828 g by 12

b) 3626 kg by 14

c) 7920 kg by 15

Worksheet - 32

Word Problems

1. A pouch had 367 g of milk powder. Of this, 125 g was used. What is the weight of the milk powder in the pouch now?

2. A container had 785 g of salt. To this, 127 g was added. What is the weight of the salt in the container now?

3. A store has 250 bags of wheat flour. Each bag weighs 12 kg. What is the total weight of flour in these bags?

4. A store has 6324 kg of cement. This is to be packed into packets of 12 kg each. How many packets can be made?

5. A godown had 4560 kg of rice. Out of this 745 kg was sold on Monday. On the next day a supply of 1032 kg was received. What is the quantity of rice in the godown now?

6. A bag full of books weighs 20 kg. 7 kg of books are taken out of this bag and 14 kg of books are added to it. What is the weight of the bag now?

Worksheet - 33

Estimating Weight

1. **Estimate the following by rounding to nearest 10 g.**

   a) 33 g ≈ _____ g   b) 759 g ≈ _____ g

2. **Estimate the following by rounding to nearest 100 g.**

   a) 256 g ≈ _____ g   b) 4195 g ≈ _____ g

3. **Estimate the following by rounding to nearest 1000 g.**

   a) 8374 g ≈ _____ g = _____ kg

   b) 5893 g ≈ _____ g = _____ kg

4. **Estimate the following by rounding to nearest 10 kg.**

   a) 357 kg ≈ _____ kg   b) 8154 kg ≈ _____ kg

5. **Estimate the following by rounding to nearest 100 kg.**

   a) 259 kg ≈ _____ kg   b) 3434 kg ≈ _____ kg

6. **Estimate the following by rounding to nearest 1000 kg.**

   a) 3874 kg ≈ _____ kg   b) 6293 kg ≈ _____ kg

7. **Solve the following:**

   a) Find the sum of 34 g and 128 g and round it to nearest 10 g.
   _____

   b) Find the difference between 245 g and 782 g and round it to nearest 100 g.
   _____

   c) Find the sum of 721 kg and 226 kg and round it to nearest 10 kg.
   _____

   d) Find the difference between 1823 kg and 2296 kg and round it to nearest 100 kg.
   _____

   e) From the sum of 5129 kg and 2305 kg subtract 1209 kg and round it to nearest 1000 kg.
   _____

Worksheet - 34

Converting Units of Capacity

Fill in the blanks.

a) 1 L = _____ ml

b) To convert L to ml we _____ by _____.

c) 3 L = _____ ml

d) 8 L = _____ ml

e) A container can hold 4 L of liquid. Its capacity in ml is _____.

f) To convert 9 L to ml we can either multiply by _____ or simply put _____ zeros in the end.

g) 1000 ml = _____ L

h) To convert ml to L we _____ by _____.

i) 5000 ml = _____ L

j) 8000 ml = _____ L

k) A tub can hold 7000 ml of water. Its capacity in L is _____.

l) To convert 6000 ml to L we can either divide by _____ or simply remove _____ zeros in the end.

m) A bottle has a capacity of 3 L and 150 ml. Its capacity in ml is _____.

n) A vessel has a capacity of _____ L and 190 ml. Its capacity in ml is 4190.

o) A vessel can hold 5 L and 700 ml of a liquid. Its capacity in ml is _____.

p) A vessel can hold 7 L and _____ ml. Its capacity in ml is 7750.

q) A water jug has a capacity of 5 L and _____ ml. Its capacity in ml is 5760.

r) A drum has a capacity of 4350 ml. Its capacity is _____ L and _____ ml.

s) A container can hold _____ L and 740 ml of juice. Its capacity in ml is 9740.

t) A generator's fuel tank when full can hold 6750 ml of diesel. Its capacity is _____ L and _____ ml.

Worksheet - 35

Adding, Subtracting, Multiplying and Dividing Capacities

1. **Add the following:**

a	b	c
5 6 ml + 2 4 ml ———— ml	3 4 7 L + 5 6 2 L ———— L	4 3 1 3 L + 4 7 1 5 L ———— L

2. **Subtract the following:**

a	b	c
7 8 ml − 2 7 ml ———— ml	4 8 3 L − 4 4 L ———— L	9 1 5 4 L − 1 8 1 9 L ———— L

3. **Multiply the following:**

a	b	c
9 7 ml × 1 1 ———— ———— ml	6 1 3 L × 1 2 ———— ———— L	3 1 7 L × 1 4 ———— ———— L

4. **Divide the following:**

a	b	c
539 ml by 11	7490 L by 14	7875 L by 15

Worksheet - 36

Word Problems

1. A beaker had 89 ml of solution. To this 85 ml more was added. What is the quantity of solution in the beaker now?
   _____
   _____

2. A container had 970 ml of oil. Of this 240 ml was used up. What is the quantity of oil in the container now?
   _____
   _____

3. A store has 761 cans of paint. Each can has 11 L of paint. What is the total quantity of paint in these cans?
   _____
   _____

4. A container has 6960 L of kerosene. This has to be distributed equally to 12 depots. What quantity is to be given to each?
   _____
   _____

5. A shop has 4120 L of kerosene oil. During the day it sold 3500 L. Then it received a supply of 2100 L. What is the quantity of kerosene the store has now?
   _____
   _____

6. A tank has 50 L of water. 22 L of water has been consumed. 15 L is now added to the water tank. What is the total quantity of water in the tank now?
   _____
   _____

Worksheet - 37

Estimating Capacity

1. **Estimate the following by rounding to nearest 10 ml:**
   a) 78 ml ≈ _____ ml
   b) 122 ml ≈ _____ ml

2. **Estimate the following by rounding to nearest 100 ml:**
   a) 586 ml ≈ _____ ml
   b) 8642 ml ≈ _____ ml = _____ L

3. **Estimate the following by rounding to nearest 1000 ml:**
   a) 5724 ml ≈ _____ ml = _____ L
   b) 6323 ml ≈ _____ ml = _____ L

4. **Estimate the following by rounding to nearest 10 L:**
   a) 83 L ≈ _____ L
   b) 169 L ≈ _____ L

5. **Estimate the following by rounding to nearest 100 L:**
   a) 461 L ≈ _____ L
   b) 524 L ≈ _____ L

6. **Estimate the following by rounding to nearest 1000 L:**
   a) 4777 L ≈ _____ L
   b) 8398 L ≈ _____ L

7. **Solve the following:**
   a) Find the sum of 264 ml and 523 ml and round it to nearest 10 ml.
   _____
   b) Find the difference between 3433 L and 7823 L. Round it to nearest 100 L.
   _____
   c) Find the sum of 916 ml, 924 ml and 960 ml and round it to nearest 1000 ml. Convert into L.
   _____
   d) Find the sum of 78 L and 34 L and round it to nearest 10 L.
   _____
   e) Find the difference between 7123 L and 2639 L. Round it to nearest 100 L.
   _____
   f) From the sum of 4143 L and 1393 L subtract 2349 L and round it to nearest 1000 L.
   _____

Unit-5 Time and Money

Worksheet - 38

Converting Units of Time

1. Write in ascending order: hour, minute, second, day, week

   _____

2. Write in descending order: century, month, decade, year, millennium

   _____

3. Fill in the blanks.

   a) 1 minute = _____ seconds.   b) 5 minutes = _____ seconds.

   c) 120 seconds = _____ minutes.   d) 1 hour = _____ minutes.

   e) 4 hours = _____ minutes.   f) 360 minutes = _____ hours.

   g) 5 hours = _____ minutes.   h) _____ hours = 600 minutes.

   i) _____ hours = _____ minutes = 7200 seconds.

   j) 1 day = _____ hours.   k) 2 days = _____ hours.

   l) 72 hours = _____ days.   m) 1 week = _____ days.

   n) 5 weeks = _____ days.   o) 21 days = _____ weeks.

   p) _____ weeks = 28 days = _____ hours.

   q) 1 year = _____ months.   r) 4 years = _____ months.

   s) 60 months = _____ years.   t) 1 decade = _____ years.

   u) 2 decades = _____ years.   v) 50 years = _____ decades.

   w) 1 century = _____ years.   x) 2 centuries = _____ years.

   y) 500 years = _____ centuries.   z) 1 millennium = _____ years.

Worksheet - 39

Time on Clocks

1. Write the correct time.

a) __:__ ____ o'clock	b) __:__ Quarter past ____	c) __:__ Half past ____
d) __:__ Three quarters past __ Quarter to ___	e) __:__	f) __:__

2. Draw the hands on the clock to show the correct time.

a) 3 : 00	b) 10 : 30	c) 7 : 05
d) 12 : 35	e) 11 : 40	f) 9 : 25

Worksheet - 40

Converting ₹ to Paise and Paise to ₹

1. **Fill in the blanks.**

    a) 7 rupees 50 paise is written as ₹ _____.

    b) 11 rupees _____ paise is written as ₹ 11.60.

    c) _____ rupees 75 paise is written as ₹ 18.75.

    d) _____ rupees _____ paise is written as ₹ 75.60.

2. **Convert the following to paise:**

    a) ₹ 57 = _____ p   b) ₹ 74.30 = _____ p

3. **Convert the following to rupees:**

    a) 3400 paise = ₹ _____   b) 8750 paise = ₹ _____

4. **Tick the smaller amount.**

    a) ₹ 24 and 30 paise ☐   b) ₹ 88.50 ☐
       2330 p ☐   ₹ 86 ☐

5. **Tick the greater amount.**

    a) 5420 p ☐   b) ₹ 67 ☐
       ₹ 54 and 25 paise ☐   6690 p ☐

    c) Arrange the following amounts in descending order:

       ₹ 58 and 25 paise, 2600 p, ₹ 25, ₹ 37.90

       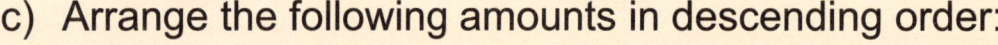

    d) Arrange the following amounts in ascending order:

       ₹ 40 and 60 paise, 4410 p, 3451 p, ₹ 53.78

       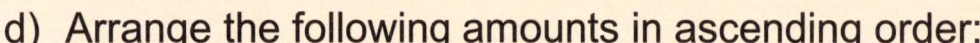

Note: 1₹ = 100 paise

Worksheet - 41

Adding, Subtracting, Multiplying and Dividing Money

1. Add the following:

a	b	c
US$ 41.32 + US$ 42.58 US$ ___ . ___	US$ 556 + US$ 444 US$ ___	US$ 5255 + US$ 2735 US$ ___

2. Subtract the following:

a	b	c
US$ 72.41 − US$ 33.65 US$ ___ . ___	US$ 353 − US$ 174 US$ ___	US$ 7965 − US$ 4737 US$ ___

3. Multiply the following:

a	b	c
US$ 3.90 × 11 US$ ___	US$ 258 × 12 US$ ___	US$ 476 × 14 US$ ___

4. Divide the following:

a	b	c
US$ 4108 ÷ 13	US$ 7154 ÷ 14	US$ 8295 ÷ 15

Worksheet - 42

Word Problems

1. Jane has to buy some gifts to be given as return gifts at her birthday party. If a gift item costs ₹ 145 what will be the cost of 13 such gifts?
   _____
   _____

2. Jennifer had ₹ 8,050 in her bank account. She withdrew ₹ 1,350 to buy a jacket. What is her bank balance now?
   _____
   _____

3. Peter had ₹ 1,365 in his purse. His father gave him ₹ 735 more. How much money does he have in his purse now?
   _____
   _____

4. John purchased a shirt for ₹ 2,250, a belt for ₹ 725 and a trouser for ₹ 2,520 from a shop. How much he spend in all? If he gave three notes of ₹ 2,000 to the shopkeeper what did he get back?
   _____
   _____

5. A prize money of ₹ 9,675 is to be equally distributed among 15 winners. How much will each get?
   _____
   _____

6. Lara buys a purse for ₹ 3,900 and a jacket for ₹ 2,450. How much has she spent in all?
   _____
   _____

Unit-6 Data Handling and Patterns

Worksheet - 43

Reading Pictographs

The following chart represents the sale of different items during a day in a garment shop.

Each point ⊙ represents one item.

Item	Point
Trouser	⊙⊙⊙⊙⊙⊙⊙⊙
Shirt	⊙⊙⊙⊙⊙⊙⊙⊙⊙⊙⊙
Cap	⊙⊙⊙⊙⊙⊙⊙
Belt	⊙⊙⊙⊙⊙
Tie	⊙⊙⊙⊙⊙⊙⊙
Jacket	⊙⊙⊙⊙⊙⊙

a) Which item has minimum sale? How many of these were sold?
_____

b) Which item has maximum sale? How many of these were sold?
_____

c) How many caps were sold?
_____

d) How many trousers were sold more than jackets?
_____

e) Sale of which two items is equal?
_____

f) How many belts were sold less than trousers?
_____

Worksheet - 44

Patterns

1. What comes next?

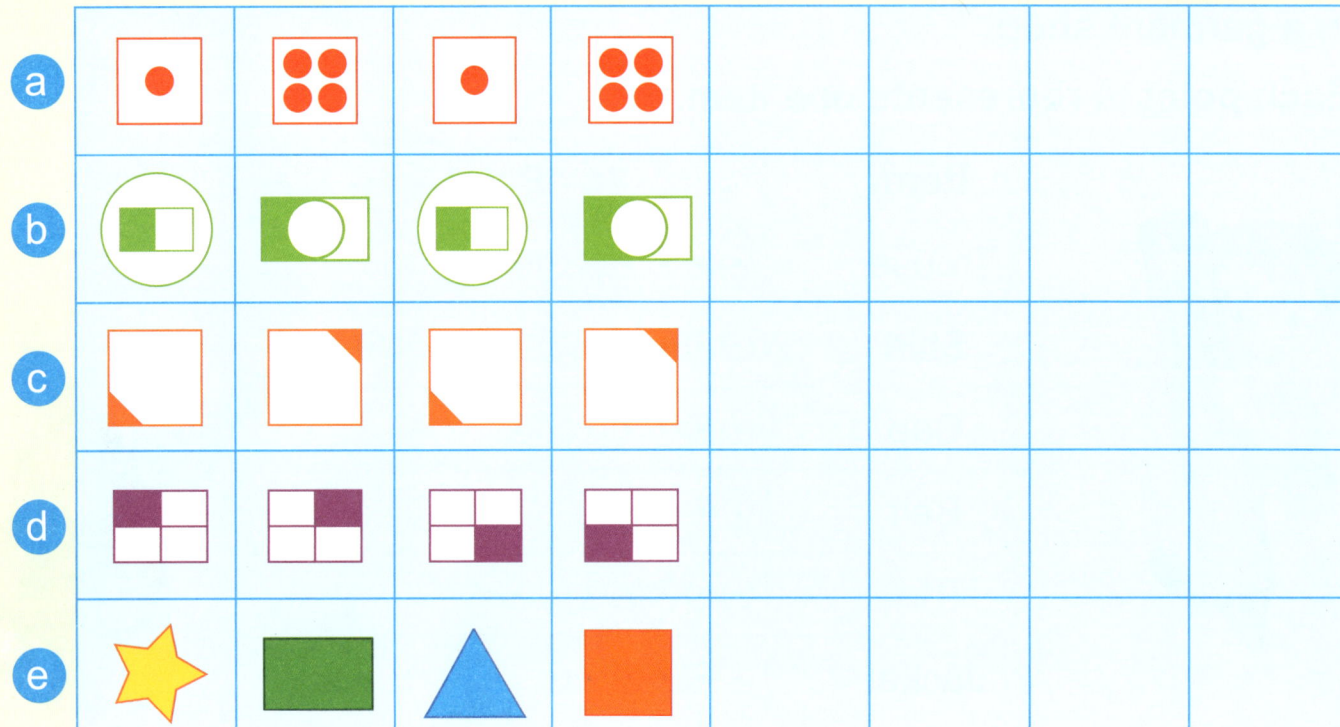

2. Fill in the blank boxes to complete the pattern.

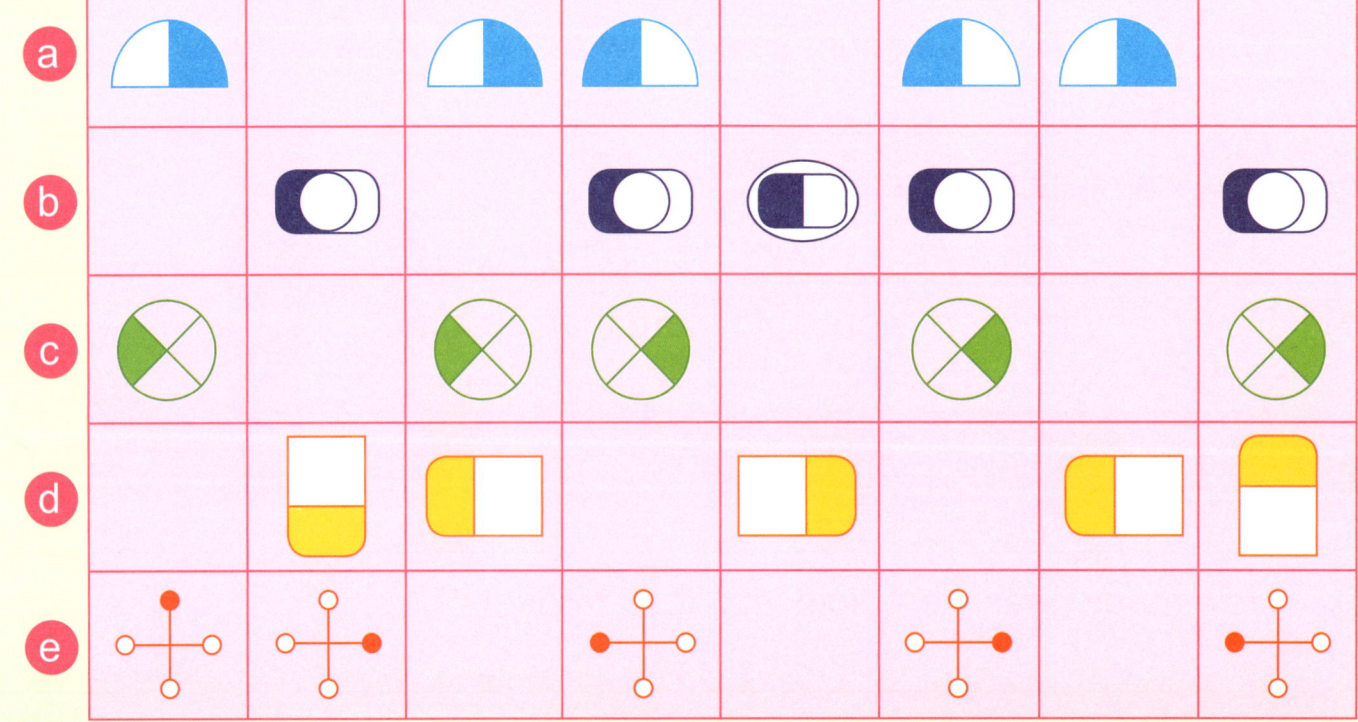

Worksheet - 45

Growing and Reducing Patterns
Numbers, Letters and Code Patterns

1. Extend the sequence.

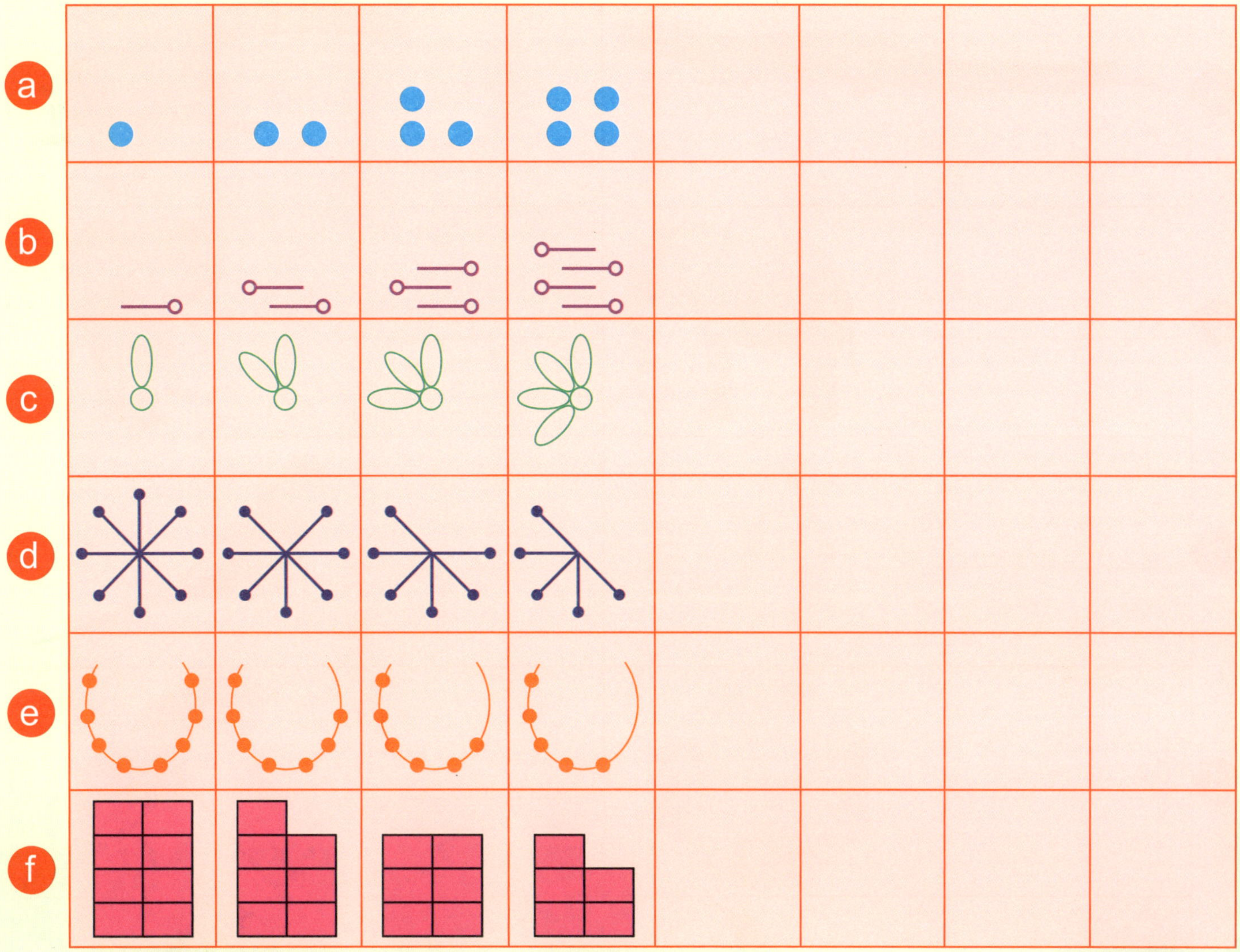

2. Complete the following number patterns:

   a) 1122, 2233, 3344, _____, _____, _____.

   b) 9999, 8888, 7777, _____, _____, _____.

   c) AZ, BY, CX, _____, _____, _____.

   d) ABC, EFG, IJK, _____, _____, _____.

   e) Z01, Y02, X03, _____, _____, _____.

   f) A1B2, C2D3, E3F4, _____, _____, _____.

Unit-7 Geometry

Worksheet - 46

Front, Side and Top View

Look at the pictures below and write which view is shown (front, side or top).

a. _____ _____

b. _____ _____

c. _____ _____

d. _____ _____

e. _____ _____

f. _____ _____ _____

Worksheet - 47

Symmetry

1. Divide the following pictures into two similar halves. If this cannot be done cross out the picture.

2. Check if the dotted line divides each figure into two similar halves. Tick mark yes or no.

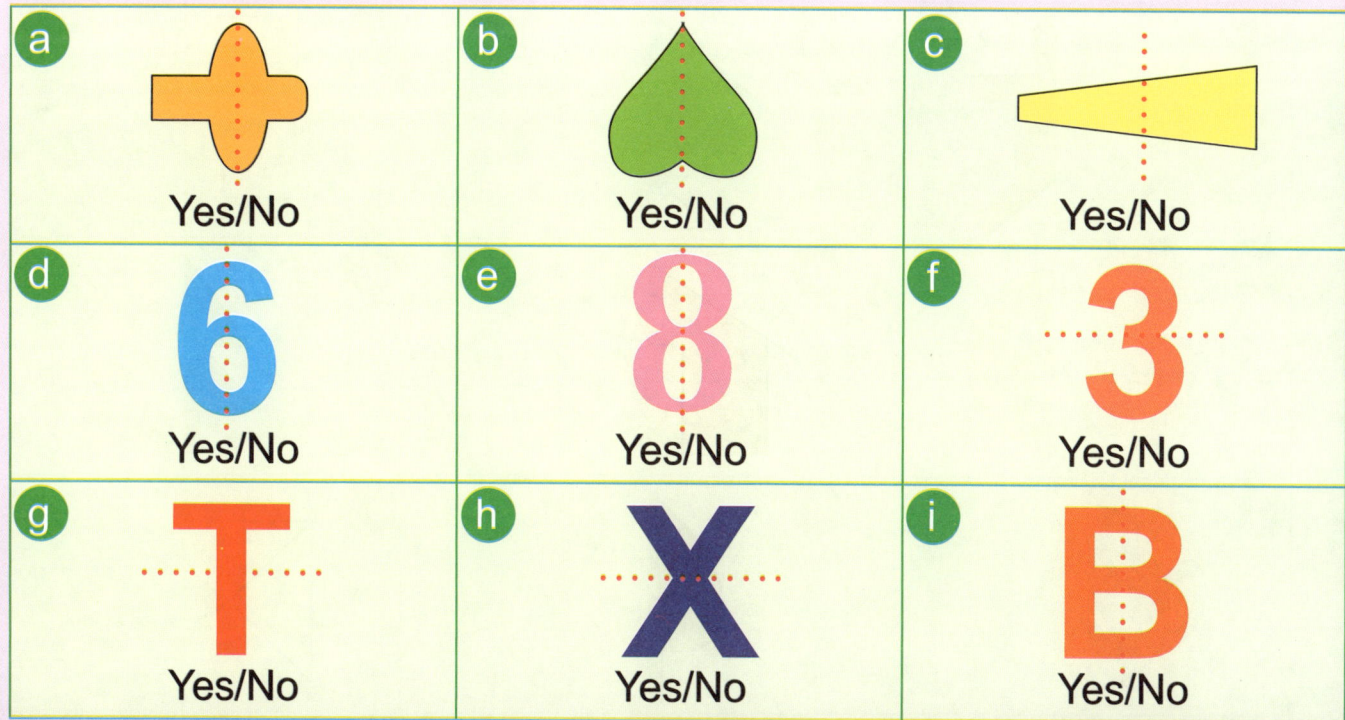

Worksheet - 48

Symmetry

1. Guess these letters from their halves. Write the complete letter in the box.

2. Taking the dotted line as the line of symmetry, complete the following pictures.

Worksheet - 49

Number of Triangles, Straight and Curved Edges

1. **Count the number of triangles in the figures given below.**

   a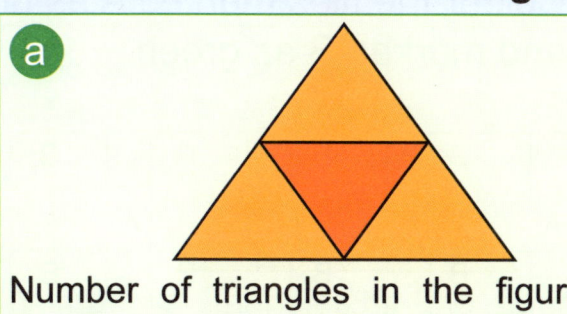
   Number of triangles in the figure are _____.

   b
   Number of triangles in the figure are _____.

2. **Fill in the blanks.**

   a
   Number of curved edge(s): _____
   Number of straight edge(s): _____

   b
   Number of curved edge(s): _____
   Number of straight edge(s): _____

   c
   Number of curved edge(s): _____
   Number of straight edge(s): _____

   d
   Number of curved edge(s): _____
   Number of straight edge(s): _____

3. **Find the number of corners, straight edges and curved edges in each of the following.**

   a
   Number of curved edge(s): _____
   Number of straight edge(s): _____
   Number of corner(s): _____

   b
   Number of curved edge(s): _____
   Number of straight edge(s): _____
   Number of corner(s): _____

Worksheet - 50

Tile Design

1. Colour the tiles marked A as red and marked B as yellow.

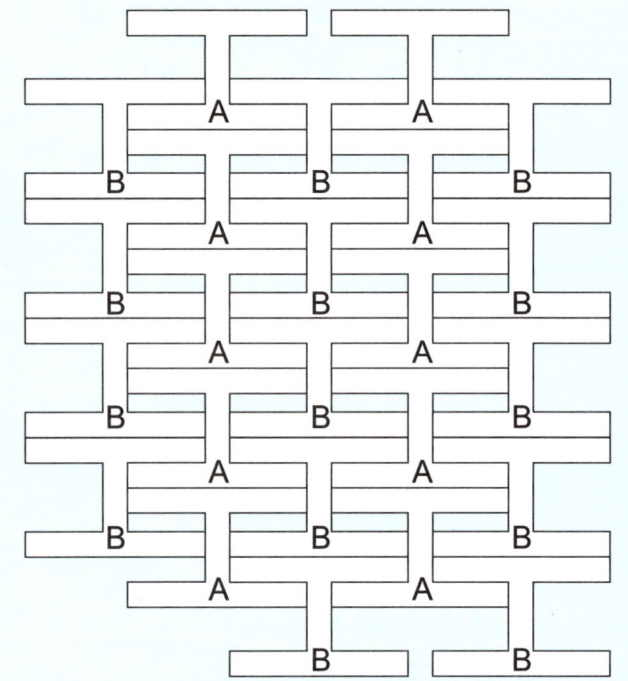

2. Colour the tiles marked A as blue and marked B as green.

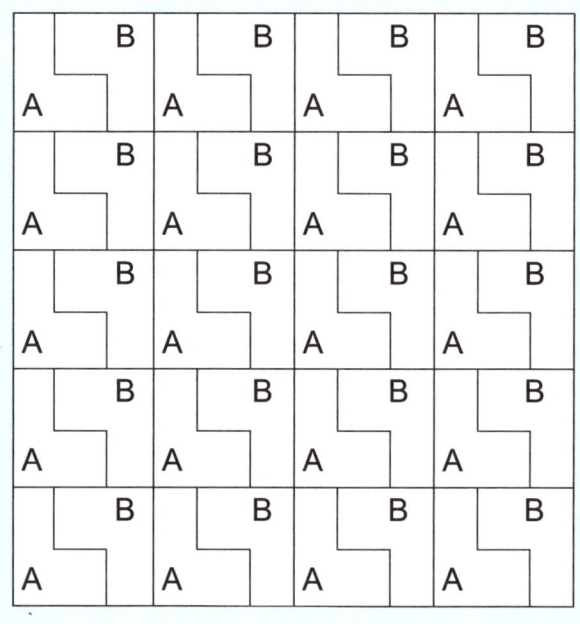

3. Colour the tiles marked A as pink and marked B as black.

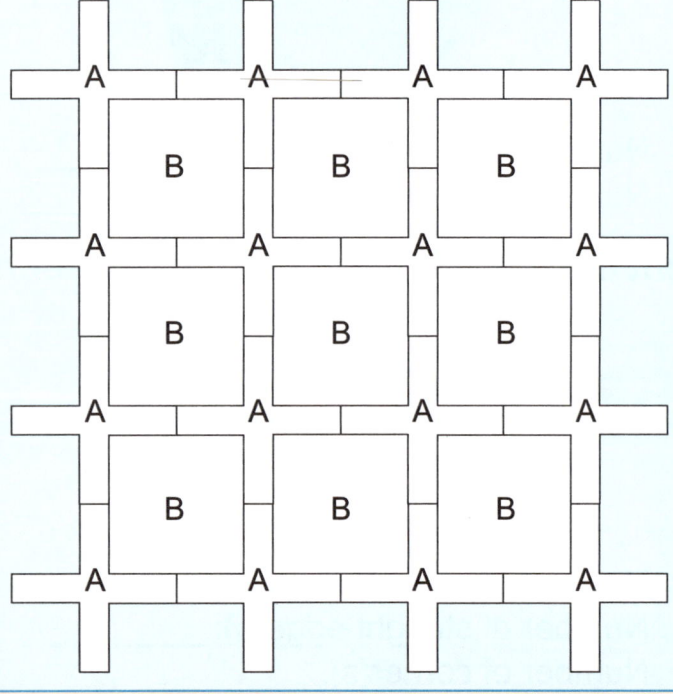

4. Colour the tiles marked A as orange and marked B as brown.

ANSWERS

Unit-1 Numbers: 1000 to 10000

Worksheet-1
1. a) 2034, 2035, 2036, 2037, 2038
   b) 4159, 4160, 4161, 4162, 4163, 4164
   c) 6403, 6404, 6405, 6406, 6407, 6408, 6409
   d) 8366, 8367, 8368, 8369, 8370, 8371, 8372
   e) 9994, 9995, 9996, 9997, 9999, 10000
2. a) 1729, 1728, 1727, 1726, 1725, 1724
   b) 3223, 3222, 3221, 3220, 3219, 3218, 3217
   c) 5203, 5202, 5201, 5200, 5199
   d) 7936, 7935, 7934, 7933, 7932
   e) 9998, 9997, 9996, 9995, 9994

Worksheet-2
1. a) One thousand one hundred twenty-one
   b) Two thousand thirty-three
   c) Three thousand three hundred fifteen
   d) Four thousand four hundred seventy-seven
   e) Five thousand five hundred forty-four
   f) Six thousand two hundred eight
   g) Seven thousand seven hundred sixty-six
   h) Eight thousand eight hundred fifty-seven
   i) Nine thousand six hundred eighty
   j) Nine thousand nine hundred ninety-nine
2. a) 1518 b) 2851 c) 3465 d) 4627 e) 5379
   f) 6432 g) 7283 h) 8246 i) 9194 j) 10000

Worksheet-3
1. a) 4259 b) 3678
2. a) 2; 3; 6; 5   b) 7; 4; 2; 9
3. a) 3231 b) 7662, 7663 c) 5698, 5699, 5700
4. a) 1236 b) 5660, 5661 c) 7997, 7998, 7999
5. a) 4123 b) 5315, 5316 c) 9997, 9998, 9999
6. a) 7270, 7272 b) 8361, 8362, 8364, 8365
   c) 9994, 9995, 9996, 9998, 9999, 10000

Worksheet-4
a) 5; 6; 8; 3   b) 6710   c) 2000; 300; 10; 7   d) 8584
e) 8000+900+20+3   f) 6763
g) (i) > (ii) = (iii) < (iv) > (v) < (vi) =
h) encyclopedia on plants
i) 4551, 8329, 8438, 9216   j) 6502, 4531, 3343, 1353

Worksheet-5
1. 9753; 3579   2. 6666; 2222   3. 8640; 4068
4. 5555; 2000   5. 7432; 1234   6. 9999; 4444
7. 9865; 5068   8. 9999; 2000

Worksheet-6
1. O's; 0; 0; T's
2. a) 4830   b) 5310   c) 6520
3. T's; 0; 0; H's
4. a) 5200   b) 4300   c) 6600
5. H's; 0; 0; Th's
6. a) 2000   b) 7000   c) 5000
7. a) 4,390   b) 4,400   c) 4,000   d) nearest ten

Unit-2 Operations on Numbers

Worksheet-7
1. a) 8957 b) 6885 c) 8678 d) 8688
   e) 8031 f) 6961 g) 7663 h) 7312
   i) 9778 j) 8987 k) 7767 l) 6896
   m) 6864 n) 8341 o) 7332 p) 6982
2. a) 7778 b) 9295 c) 6182 d) 8180
   e) 5716 f) 8055

Worksheet-8
1. 8,633   2. 6,166   3. 3,601
4. 4,186   5. 5,825   6. 3,023

Worksheet-9
1. a) 5400   b) 5910   2. 8,330
3. a) 5400   b) 7400   4. 8,900
5. a) 9000   b) 8000   6. 9,000
7. a) 7123   b) 3874; 4623   c) 2655
   d) 2563; 4359; 3178   e) 0
   f) 4295   g) 0

Worksheet-10
1. a) 2412 b) 3321 c) 4333 d) 4565
   e) 2314 f) 2040 g) 3342 h) 3255
   i) 1227 j) 3191 k) 2701 l) 3087
   m) 4887 n) 1288 o) 2129 p) 2568
2. a) 3757 b) 5651 c) 1122 d) 3596
   e) 2364 f) 7656 g) 1889 h) 7589

Worksheet-11
1. 40   2. 5821   3. 4313   4. 2643
5. 1665   6. 4517   7. 1945   8. 5777

Worksheet-12
1. 5,235   2. 2,204   3. 168
4. 1,083   5. Sam; 664   6. 2,290

Worksheet-13
1. a) 4480   b) 2880   2. 1,820
3. a) 2800   b) 2800   4. 1,300
5. a) 3000   b) 4000   6. 2,000

Worksheet-14
1. a) incorrect   b) correct
2. a) correct   b) incorrect
3. a) 3424   b) 0   c) 3952
   d) 2475   e) 3677   f) 0

Worksheet-15
1.
11	12	13	14	15
22	24	26	28	30
33	36	39	42	45
44	48	52	56	60
55	60	65	70	75
66	72	78	84	90
77	84	91	98	105
88	96	104	112	120
99	108	117	126	135
110	120	130	140	150

2. 88   3. 48   4. 117   5. 84   6. 105

Worksheet-16
1. 8534   2. 9288   3. 8552   4. 7820
5. 7338   6. 8659   7. 9176   8. 9477
9. 6876   10. 5902   11. 7616   12. 8055
13. 9270   14. 6902   15. 8892   16. 7044

Worksheet-17
1. 2,585   2. 7,500   3. 4,888
4. 8,008   5. 8,775   6. 480

Worksheet-18
a) 4312   b) 2108   c) 1232; 5395
d) 3982   e) 4991; 3642
f) 4139; 5126; 3563   g) 1   h) 2616
i) 1   j) 5116   k) 0   l) 0
m) 0   n) 0   o) 4570   p) 9380
q) 270   r) 2700   s) 7200   t) 4,200
u) 2000   v) 7000   w) 8,000

Worksheet-19
1. a) 4680   b) 3960   c) 8,960   d) 7200
   e) 9500   f) 7,400   g) 8000   h) 6000
   i) 9,000   j) 4800   k) 2100   l) 3,200
   m) 8000   n) 6000   o) 8,000   p) 8400
   q) 9600   r) 6,900
2. a) 3000   b) 3900   c) 4,000

Worksheet-20
1. 6   2. 8   3. 5   4. 48   5. 62
6. 38;10   7. 74; 7   8. 43; 7   9. 63; 11

Worksheet-21
1. 649   2. 732   3. 521   4. 356   5. 325
6. 616; 6   7. 416; 3   8. 231; 5   9. 473; 1

Worksheet-22
1. 7   2. 74; 3   3. 364   4. 482; 10

Worksheet-23
1. a) 347   b) 734   c) 675   d) 34   e) 46
   f) 51   g) 5   h) 8   i) 9
2. a) 1   b) 1568   c) 6393   d) 0
   e) 0   f) 381; 12   g) 6773; 13
3. a) 7   b) 3

Unit-3 Fractions

Worksheet-24
1.
2.
3.
4.
5.
6.
7.

Worksheet-25
1. 3/4   2. 2/3   3. 1/4   4. 1/5
5. 1/2   6. 2/5   7. 1/3   8. 3/7
9. 1/7   10. 3/8   11. 8/9   12. 3/10

Unit-4 Measurement

Worksheet-26
a) 100 b) 400 c) 1200 d) 1
e) 4 f) 34 g) 1000 h) 8000
i) 7000 j) 1 k) 2 l) 5
m) 1715 n) 29; 50 o) 1270 p) 67; 25
q) 7200 r) 9; 400 s) 7625 t) 5; 720

Worksheet-27
1. a) 98 b) 782 c) 9994
2. a) 55 b) 596 c) 1782
3. a) 803 b) 1740 c) 5356
4. a) 34 cm b) 157 m c) 587 km

Worksheet-28
1. 577 m 2. 1132 m 3. 2772 m
4. 6 m 82 cm 5. 402 m 6. 101 m

Worksheet-29
1. a) 80 b) 140 2. a) 600, 6 b) 900, 9
3. a) 50 b) 180 4. a) 900 b) 7400
5. a) 7000, 7 b) 3000, 3 6. a) 900 b) 1210
7. a) 900 b) 2400 8. a) 3000 b) 7000

Worksheet-30
a) 1000 b) multiply; 1000 c) 4000
d) 8000 e) 5000 f) 1000; three
g) 1 h) divide; 1000 i) 3
j) 8 k) 7 l) 1000; three
m) 6250 n) 560 o) 4
p) 4; 510 q) 2350 r) 7
s) 720 t) 6; 830

Worksheet-31
1. a) 78 b) 887 c) 8231
2. a) 12 b) 6662 c) 2568
3. a) 4668 b) 2808 c) 6076
4. a) 69 g b) 259 kg c) 528 kg

Worksheet-32
1. 242 g 2. 912 g 3. 3000 kg
4. 527 5. 4847 kg 6. 27 kg

Worksheet-33
1. a) 30 b) 760 2. a) 300 b) 4200
3. a) 8000, 8 b) 6000, 6 4. a) 360 b) 8150
5. a) 300 b) 3400 6. a) 4000 b) 6000
7. a) 160 g b) 500 g c) 950 kg
   d) 500 kg e) 6000 kg

Worksheet-34
a) 1000 b) multiply; 1000 c) 3000
d) 8000 e) 4000 f) 1000; three
g) 1 h) divide; 1000 i) 5
j) 8 k) 7 l) 1000; three
m) 3150 n) 4 o) 5700
p) 750 q) 760 r) 4; 350
s) 9 t) 6; 750

Worksheet-35
1. a) 80 b) 909 c) 9028
2. a) 51 b) 439 c) 7335
3. a) 1067 b) 7356 c) 4438
4. a) 49 ml b) 535 L c) 525 L

Worksheet-36
1. 174 ml 2. 730 ml 3. 8371 L
4. 580 L 5. 2720 L 6. 43 L

Worksheet-37
1. a) 80 b) 120 2. a) 600 b) 8600; 8.6
3. a) 6000; 6 b) 6000; 6 4. a) 80 b) 170
5. a) 500 b) 500 6. a) 5000 b) 8000
7. a) 790 ml b) 4400 L c) 3000 ml, 3 L
   d) 110 L e) 4500 L f) 3000 L

Unit-5 Time and Money

Worksheet-38
1. second, minute, hour, day, week
2. millennium, century, decade, year, month
3. a) 60 b) 300 c) 2 d) 60 e) 240
   f) 6 g) 300 h) 10 i) 2; 120 j) 24
   k) 48 l) 3 m) 7 n) 35 o) 3
   p) 4; 672 q) 12 r) 48 s) 5 t) 10
   u) 20 v) 5 w) 100 x) 200 y) 5
   z) 1000

Worksheet-39
1. a) 5:00, 5 b) 8:15, 8 c) 11:30, 11
   d) 11:45,11,12 e) 9:05 f) 2:25

2. a)  b)  c)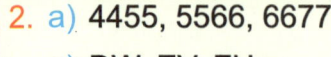
   d)  e)  f)

2. a) 4455, 5566, 6677   b) 6666, 5555, 4444
   c) DW, EV, FU   d) MNO, QRS, UVW
   e) W04, V05, U06   f) G4H5, I5J6, K6L7

Unit-7 Geometry

Worksheet-46
a) side; front   b) side; front   c) front; side
d) top; side   e) front; side   f) side; front; top

Worksheet-40
1. a) 7.50   b) 60   c) 18   d) 75; 60
2. a) 5700 b) 7430   3. a) 34   b) 87.50
4. a) 2330 p   b) ₹ 86
5. a) ₹ 54 and 25 paise   b) ₹ 67
   c) ₹ 58 and 25 paise, ₹ 37.90, 2600 p, ₹ 25
   d) 3451 paise, ₹ 40 and 60 paise, 4410 p, ₹ 53.78

Worksheet-41
1. a) 83.90   b) 1000   c) 7990
2. a) 38.76   b) 179   c) 3228
3. a) 42.90   b) 3096   c) 6664
4. a) US$ 316   b) US$ 511   c) US$ 553

Worksheet-42
1. ₹ 1,885   2. ₹ 6,700   3. ₹ 2,100
4. ₹ 5,495; ₹ 505   5. ₹ 645   6. ₹ 6,350

Unit-6 Data Handling and Patterns

Worksheet-43
a) Belt, 5   b) Shirt, 11   c) 7
d) 2   e) Cap, Tie   f) 3

Worksheet-47
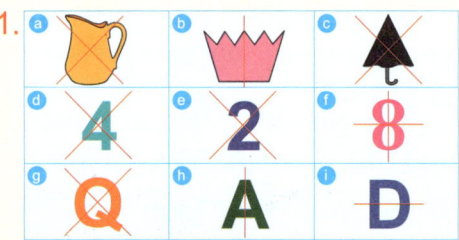
2. a) No   b) Yes
   c) No   d) No
   e) Yes f) Yes
   g) No   h) Yes
   i) No

Worksheet-48
1. a) X   b) T   c) H   d) C   e) U
   f) A   g) M   h) W   i) B

2.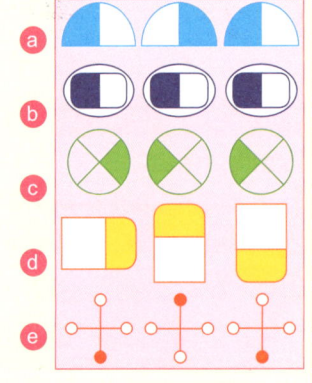

Worksheet-44
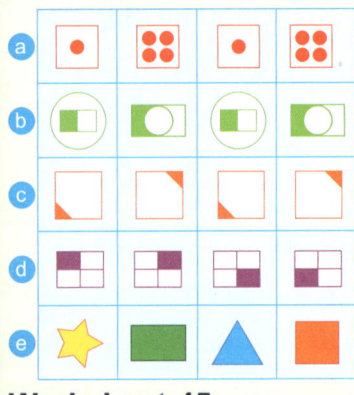

Worksheet-49
1. a) 5   b) 8   2. a) 1; 0   b) 0; 12
   c) 2; 2   d) 2; 10   3. a) 2; 0; 2   b) 2; 2; 4

Worksheet-50
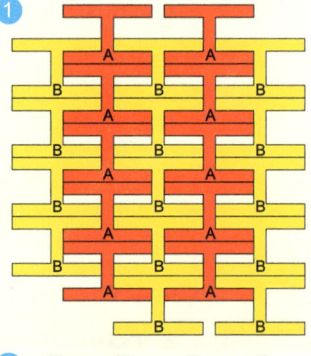 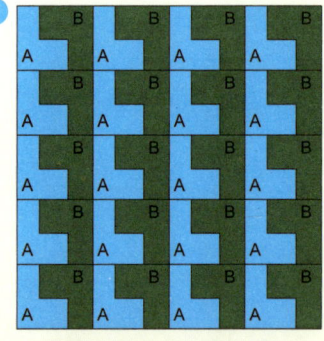
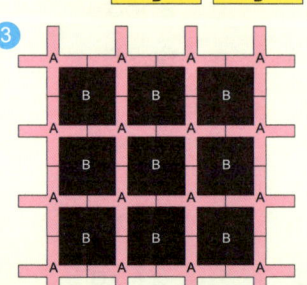 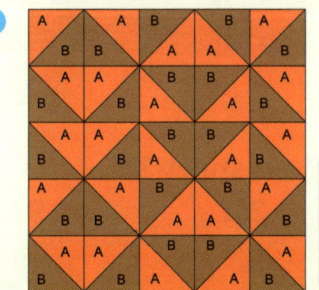

Worksheet-45
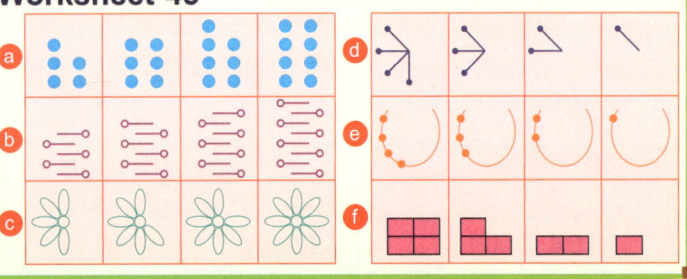

56